To the Robinsons—
Love from Lacey + David
Helena, Montana

January 2005
XOXO

Montana
WILDLIFE PORTFOLIO

FARCOUNTRY
PRESS

Photography of **DONALD M. JONES**
Foreword by **RICK BASS**

RIGHT: Bull elk poised against the Mission Mountains and the dawn of a new day on the National Bison Range, Moiese.

PAGE ONE: Sandhill cranes against a Montana sunrise.

FRONT COVER: A grizzly takes a mid-day soaking in Yellowstone national Park.

BACK COVER (TOP LEFT): Leaping sandhill crane.

BACK COVER (TOP RIGHT): A red fox kit in a stare-down.

BACK COVER (BOTTOM): A bull elk races in pursuit of his cows.

ISBN: 1-56037-240-0

Photographs © Donald M. Jones

© 2003 Farcountry Press

For more information on our books write: Farcountry Press, P.O. Box 5630, Helena, MT 59604 or call: (800) 821-3874, or visit www.montanamagazine.com

Created, produced, and designed in the United States.
Printed in Korea.

Dedication

To Tess, Jake and Luke.
Thank you for all your support and understanding
in allowing me to pursue my passion.
I love you.

FOREWORD

by Rick Bass

Up at the crack of dawn, a whitetailed buck.

Admit it: there's rarely a day that goes by—if you're not blessed to live in Montana—that you don't wish you were. I'm overwhelmingly grateful for the privilege of living here, and feel sometimes as if there's been some mistake, and that someone, somewhere else, is paying all the dues, while all the luck flows this way. I feel that way every time I step outside, in any weather, and I feel that way, looking at these pictures.

With the great fortune of having Don Jones for a neighbor, I've seen his work before, even some of the slides reproduced in this book. Although these images are from all around the state of Montana, many of the landscapes are deeply familiar to me, coming as they do from Don's and my beloved Cabinet-Yaak ecosystem, and it occurs to me that I might even have seen one or two of the individual subjects, our paths crossing and weaving in that mysterious and unknowable, unseen braid that informs the spirit of any place— though I am aware, too, looking at Jones' work here, Jones' craft and calling, that I am looking at ghosts and shadows; that even for those of us who take great delight in trouncing around in the backcountry, or even slipping around fairly quietly sometimes, Jones is a far more enthusiastic, and yet focused, traveler out there, maddeningly patient—possessing, indeed, the patience of a saint.

For those of us privileged enough to spy a track, or hear a distant elk bugle, or examine an antler-torn sapling, or spy a mountain lion stalking a mule deer, or even spot a distant bear hurrying away from us—a flash of fur, a swirl of wing, a ribbon or tendril of scent, then nothing, gone, only memory—these images of dignity and grace are like the anchors or objects that cast those shadows. We're able to examine in detail how it was, and how it is. What we saw or heard or scented was so often only the shadow of these animals, the shadows thrown by their movement. It occurs to me too that these photographs can be viewed as almost a kind of dreamland: not Jones' or that of the animals, but our own; that on the mornings when we slept in until daylight, or came in from out of the rain or snow, or made the town run for errands—while Jones remained out in the sog and slosh of the cattails, battling blizzards or mosquitoes or capricious light conditions—we chose to inhabit a shadow world, away from the hopefully-enduring real world depicted in these pages: a world of muscle and bone, antler and stone and feather, claw and hoof and creek and cloud.

Two things Jones does not have patience for are digital enhancement—a technological regression that threatens the continuing existence of his craft as surely as fragmentation and loss of habitat often threaten his wild subjects—and the photographic use of captive animals: technicolored livestock, prisoners, whose movements and expressions often may not indicate any more about the true nature of their species, and the country the world really desires for that species to inhabit, than would a diamond ring inform a viewer about the true character, true heart, of the man or woman wearing it.

Jones' photographs inform us, in these blinking moments when we step through his shutter from the dream world into the real world, not just about individuals, but also about their habitats. This is what white-tailed deer do in the rain when they're too wet, this is the kind of slope mule deer bed on, sandwich-

Coyote pups ready to go out and play.

ing themselves in midslope to catch cooling evening's downward-drifting air currents and scents from the ridge above, while visually scanning the downwind slope below. Jones' photographs tell stories about individual wild animals, as well as their habitats, in ways that photographers (and viewers) of captive stock will likely never see or know. The kestrel amidst the September-dead browned cottonwood leaves, clearly a drought year, looking trim and, who knows, brooding on the impending migration, surprised, perhaps, by the photographer beneath its perch.

The buzzing of the flies in the sun's warmth, a warmth made sweeter by that of an elk's flank: a balanced world of sound and odor, as well as sight.

Nature's patterns and designs are other hallmarks of a Don Jones photograph, with those ancient patterns and designs the outcome of the relentless hand of a loving world seeking to sculpt the tangents of those fits and patterns. The battling elk whose antler growth follows the branching patterns of the trees in the forest in which it now hides, antlers sculpted also as if to perfectly encage one's opponent without necessarily puncturing or goring that opponent. The crested little feathers on the head of the common redpoll, so similar to the dried sunflower seedhead on which the bird rests, provides only one brief sense of grace in a nearly infinite number of moments that are present in the world, waiting to be observed.

There it is again, always again, and Jones finds it, and brings it to us—like accepting like, compromise and fit carving and hammering themselves out everywhere in the great and often secret society of the wilderness.

Not just shape and texture, but tone and color, too, are part of these patterns—the coyote pups the color of the soil in which their burrow is dug, the deer the tone of the grass through which they swim—and it is this composition detailing the relationship between the individual and its environment that again and again seals the authenticity, the integrity, of a Don Jones photograph, and instructs the viewer while honoring the subject.

Through his portraits of both individuals and populations, Jones captures Montana's biological wildness, rather than the more commonly portrayed recreational wildness. It is refreshing to me that it is rarely the antlers one remembers from the pictures of Jones' photos of deer and elk, but rather the ecological instruction of those wild portraits. This is how a mule deer buck—not a tame or human-conditioned one, but a backcountry deer, with antlers autumn-shined—responds to being surprised; this is the rocky country that has created the seed, idea, then motion of such a creature, a shape and movement so wedded to place that it seems certain such wedding must always have existed, at least in a dream, a dream within that rock or clay or dust.

Another Jones lesson, I think, concerns the confederacy of families—mother, young, wild siblings—and another classic strength of Jones' work is the gaze of the animals, a gaze he meets and greets with a dignity and respect uncommon in the world, but particularly so in the distance between animals and humans. The bears looking downslope with uncertainty, in a way that suggests the wind is changing beneath their reef; the reverie of the summer-fat squirrel

A Clark's nutcracker sits precariously on a spruce bow. Later named after Captain William Clark, the species was first sighted by him on August 22, 1805, in future Montana.

plump in bright kinnickinnick berries while gorging on pine nuts. Slowly, cumulatively, it is the gaze of the animals one begins to notice, and it is that rarest of things, the gaze—and story—of wild and free and uncompromised individuals...

There is a generosity of spirit in Jones that extends beyond his work, present also in his relations with family and community, as well as landscape. Montana is fortunate to have him, surely helps shape him as it shapes all of us, and perhaps in some way he is even helping to shape not just us, with his instruction and observation—his notice—but even the land itself, through the lens of celebration. Encouraging us to pay attention to, and respect, the rank flowing fecundity—not just the glorious rock and ice of the state, but the other spectacle, the phenomenon of life.

At the risk of insulting the photographer, I want to suggest that Don Jones has a lot of luck, in addition to the obvious skill and experience. By this I mean surely he's paid his dues with years of unnecessary bad luck—a staggering calendar of should-have-been-here-yesterdays. Part of me wants to believe that that photograph of the rut-wild bull rolling his eyes in the snowstorm was such luck, and that it just happens to Don again and again. I believe, however, the old adage that luck is the residue of preparation. As a neighbor of Don's here in the heart of the Cabinet-Yaak, and as a Montanan, it is my fervent hope that he will keep on putting in all that extra, necessary preparation, while the rest of us reap the benefits of his work, and receive an education about our home, in these lessons that are most pleasing to the eye, the mind, and the heart.

A bull elk races in pursuit of his cows.

Though not a significant part of their diet, grizzly bears like this one at Logan Pass, Glacier National Park, hunt ground squirrels, spending a great deal of time and energy for what seems like nothing more than a snack to such a large animal.

These bachelor mule deer, caught in a May snow squall, will probably remain together for the entire summer, but as fall approaches and the velvet sheds, their groups will break up, with the mature bucks establishing their territories before the rut.

Napping in a bed of horsetail, a bull elk enjoys summer's warm days, with food a-plenty and no territory to defend or snow to plow through. Even while he sleeps, his ears are always at work, like miniature radars in search of lurking danger.

A great gray owl flies silently on it's 4 ½-foot wing span through the Kootenai National Forest.
This largest owl in North America lives primarily on mice and voles.

ABOVE: Although he looks to be bellowing, this bull moose is actually lip-curling, or flemen, opening the entrance to the Jacobson's organ located on his upper palate to detect if a cow is in estrus.

FACING PAGE: Barley fields adjacent to Freezeout Wildlife Management Area along Montana's front range attract migrating snow geese, up to 500,000 of which will stop here on their way to and from their nesting areas in the Canadian Arctic.

This three-week-old coyote pup shares a den with parents and eight siblings under a fallen tree in the Kootenai National Forest. The cubs' days seem to be filled with playing and exploring as they wait for their parents to return with the day's catch.

LEFT: Reserved seating only! It's summer, and if you're this western painted turtle, there's only one large rock in the pond before you learn to get around.

BELOW: Although not native to the United States, ring-necked pheasants like this rooster have established themselves quite well in Montana, where they're hunted by locals and visitors each autumn.

ABOVE: Late spring snow on St. Marys Peak keeps an elk herd in Glacier National Park from embarking on journeys that will take them from their wintering grounds on the Front Range to their summer range higher in the Rocky Mountains.

FACING PAGE: A large female grizzly—her two cubs just out of sight—pauses on a rock allowing my camera's motor-drive to whirl. I am standing on the road, and will stand nowhere else when filming these great creatures. I believe they know that the road belongs to me, but that everything off the road is theirs.

Pronghorn bucks like this one on the National Bison Range will shed their horn sheaths starting in the middle of October each year, leaving bony blades on their heads until they regrow the
sheaths. By the age of three, most bucks have horns between twelve and thirteen inches long.
Bucks between the age of four and five are said to be in their prime.

LEFT: The smallest falcon in North America, the sparrow hawk, or kestrel, is found throughout Montana. It feeds primarily on small rodents and birds and, when larger insects like grasshoppers are available during the summer, one can find these small falcons hopping on and off fence lines in pursuit.

BELOW: Sage grouse are Montana's largest members of the grouse family, weighing in at more than six pounds for a healthy male. These grouse are taking part in the spring mating ritual of dancing and fighting to impress available hens. The gathering site, called a lek, is used year after year, sometimes for decades.

LEFT: Black bear cub-of-the-year playing with a willow sapling. Born in a den sometime in January, it is now about four months old. Spring days are for exploration and playing with its sibling, while depending on mother for milk. As summer progresses, playing time will dwindle and learning to eat will take precedence.

BELOW: A sure sign of spring in western Montana is the sound of the ruffed grouse's "drumming," in hardwood thickets. This bird's beating his wings from atop his favorite log assures all that spring is in and winter is out.

Ouch! Finding this ram acting as a pincushion has to be one of my more perplexing witnesses in nature. How and why it has these quills, I do not know, but they did not seem to slow him down any as he fed along with other rams. None of the quills was in a vital area, but whether they would work their way in deeper was anyone's guess.

A male red fox is greeted by his kits. Often, when an adult returns to the den, it is greeted by its kits' licks to the mouth, hoping for a reward of some regurgitated feast—yummm, sounds good.

In late August, whitetailed deer bucks' antlers, though still in velvet, are fully developed. Over the next couple of weeks the velvet will give way to hardening horn, and all the animals' coats will change from the reds of summer (as on the closest buck) to the grays of fall and winter (back two bucks).

RIGHT: The pine marten is just one of several members of the weasel family found in Montana. This particular individual was frequenting a bird feeder at a friend's home. Everything was going fine until my friend made his presence known, and the marten ran away. They came to an arrangement, though, and the marten stays away from the feeder if my friend throws out a couple of pieces of dog food for him.

BELOW: Montana is part of the Central Flyway of North American waterfowl. Each spring hundreds of thousands of ducks—like these mallards and pintails—make their way through the state, particularly the central plains and Rocky Mountain front.

FACING PAGE: The massive antlers of this bull elk serve not only as protection in his fight dominance during the rut, but also make great back scratchers—as this photo shows. I have even seen bulls use their antlers as pruning shears to bring down limbs so they can eat the succulent tips.

ABOVE: Long-legged American avocets, like this one on the central Montana prairie, use their upturned bills to capture small insects and invertebrates. They can be found hunting by swinging their bills from side to side in prairie potholes and lakes throughout Montana.

FACING PAGE: An American bison bull with hitchhiking cowbird in the National Bison Range, which was established in 1908 by the American Bison Society with approximately forty-seven bison but now boasts a healthy population of between 350 and 500 animals.

Out of sight down below these two mule deer bucks, a lone doe feeds while both bucks keep equal distance from her. Once she comes into estrus, the truth will be known as to which buck is dominant.

Sensing danger, this female grizzly takes to her back legs for a better view. I don't think there is a more on-edge animal walking the planet than a female grizzly with this-year's cubs. And reasonably so, since two thirds or more of all grizzly cubs do not make it through their first year. Five days after I took this photograph, this female had only one cub with her.

The Canada goose has to be Montana's most prevalent waterfowl. As in most of the

Sounds of September ring out from this aspen stand. It's late in the month, and the elk rut is in full swing. This fellow's problem is that he is not the area's dominant bull, and therefore must keep his distance from the others—or be forced to use that headgear.

Western Montana is home to a fairly large moose population, especially in the northwest. Of the three sub-varieties of moose, the Shiras moose lives here.

Double-crested cormorant in breeding plumage. These birds are most frequently found along Montana's major rivers and reservoirs, where they feed primarily on fish, and breed in colonies where one tree may have dozens of nests. Because they use their nests year after year, their trees stand out, white from years of accumulated excrement.

LEFT: Don't bother hanging on, help isn't on the way! I found two four-month-old black bear cubs playing in a small lodgepole pine. One climbed out onto a small limb and started to fall, holding on only by its claws or teeth, then its supportive sibling would bite its paw or nose until the first one fell. This went on for about twenty minutes.

BELOW: Breakfast in bed. A red-necked grebe feeds a newborn while the mother incubates the last egg. Montana is home to six species of grebes, all of which build floating nests of reeds, like this one, on ponds and lakes.

Whitetailed deer once were found primarily in the eastern part of Montana, but now are well established to all four corners of the state. Where their ranges overlap with those of mule deer, hybridization may occur.

A coyote sits out an impending storm before resuming food scavenging—for carrion, mice, or birds. Coyotes' adaption is remarkable, as proven by their eastward expansion across the United States.

A northern pygmy owl watching a flock of pine siskins. The northern pygmy may be Montana's smallest owl, at 6 ½ inches tall, but it is also the most aggressive: not uncommonly they take down birds twice their size, such as robins or mourning doves.

Of all the bears I film, the sub-adults, like this grizzly, seem to be the most cantankerous—grizzlies from two to four years old, and black bears that are two to three. They tend to be wanderers without established territories, so adult bears bounce them out of almost everywhere they go.

It's mid-August and the velvet starts to shed. It's not a painful process, despite the bleeding. The whole process is usually pretty rapid, taking about twenty-four hours from the start to finish. Sometimes, but not always, bull elk eat the velvet.

This coyote den was located near a road, allowing the pups to get fairly
acclimated to human onlookers. Coyotes and foxes are prone to move
their dens several times while with young, and this den was moved away

RIGHT: I often rely on friends for information on subjects to photograph, and this calliope hummingbird family was one of those cases. Searching out a hummingbird nest is like finding a needle in a haystack, but here I was able to film the young from eggs to fledglings.

BELOW: Members of the weasel family have very high metabolisms and therefore must eat a lot and often. This mink took a closer look at me, but found me interesting only for a moment. It's like "If I can't eat you, I don't have time to stick around."

A quick shake, and this whitetailed buck is as good as new. Photographing in the rain is not usually fun, but sometimes it can catch unusual behavior. Without the rain, this buck would have gotten up from his bed and just moved away, but he really enjoyed five seconds of shaking off the moisture.

What a delicious September sunbath for a large group of pronghorn does and fawns! During their rut, pronghorn males may acquire up to twenty or more does/fawns in their harems, and often the bucks sequester them in coulees or other secure sites—while keeping other bucks at bay.

"Looks bad, Bob, real bad!" I've had the good fortune to have been able to work at a fox den in the Missoula area over about five years. For two of those years, I discovered the den supported two different litters. Here are two kits that are several weeks apart in age.

ABOVE: This red-tailed hawk's nest sits atop an eighty-foot Douglas-fir, so I photographed it from a makeshift blind on an 87-foot cliff. As the bird held a yellow racer in its beak, I thought it knew I was there, but let me be "out of sight, out of mind."

RIGHT: Pacific chorus (tree) frogs are rather prevalent in Montana's northwest. Their call echoes through the woods during the later spring days, as they search for mates.

A brown-phase black bear gives his impression of a chaise lounge. He had been down on all fours when he suddenly got an itch on his right side, and threw himself on his back and began to scratch his right side with his right paw (no kidding).

A bull elk presides over his harem in the Flathead Valley.

Although born with spotted coats in late May to early June, by autumn—
as seen here—elk calves look like miniatures of their mothers.

ABOVE: A male western bluebird searches for insects upon a lichen-covered rock. Besides this species, Montana is also home to eastern and mountain bluebirds.

RIGHT: Mountain goats that gather on rocky ledges around Glacier National Park's Logan Pass, like this nanny, give human visitors a marvelous viewing opportunity.

Rest for a mule deer buck on a fog-laden ridge.

Taking the plunge for a meal. A coyote's hearing is so acute that
it may hear a mouse or vole beneath eighteen inches of snow.

ABOVE: Great horned owls are among Montana's earliest nesters, with incubation starting even in February. This one finds refuge during daylight hours on a north-facing rocky cliff.

RIGHT: The prairie rattlesnake is Montana's only poisonous snake—but fear little, as it generally is not aggressive, and tries to avoid contact with humans.

Don't even entertain the idea of trying to cross the river and seduce this bull elk's cows.

Willow snack for a bull moose during a December snowstorm. Because of their long legs and large chests, they must get down on their knees if they want to eat something on the forest floor.

RIGHT: This bald eagle is having kokanee salmon for dinner. Bald eagle numbers in Montana have grown tremendously over the past ten to fifteen years, with Canyon Ferry Reservoir on the Missouri River near Helena becoming one of the better places for winter viewing of the magnificent birds.

BELOW: Hoary marmots go head to head. These marmots are found mostly in Montana's alpine areas, with their cousins, yellow-bellied marmots, preferring lower elevations.

A gray wolf trudges through deep snow after a unsuccessful run on a cow elk.
After having many encounters with wolves over sixteen years, opportunity
finally knocked, allowing me to document wild wolves at close range.

Despite being very wide-ranging across Montana, mule deer (like this one by an old-growth ponderosa pine) are seeing their numbers shrink in the state, while whitetailed deer are on the increase.

Most autumn skirmishes between bull elk last but a few seconds, but some may go on for several minutes, and may even result in death to one of the combatants.

Though gangly, this two-year-old grizzly bear, if a male, may reach more than 600 pounds and stand above seven feet tall in its maturity.

61

ABOVE: Tall, cool grass makes a perfect refuge for a whitetailed buck on a hot July day.

RIGHT: A harrier's (marsh) hawk nest—with four chicks—lies hidden in a prairie-pothole marsh.

Upon returning to the den, this adult coyote is greeted by a hungry youngster. Coyotes are one of four species of canine on the Montana landscape, with swift fox, red fox, and wolf.

ABOVE When bighorn rams' horns curl around so that the tip meets the animal's eye, they are said to be in "full curl." A ram is about eight years old by then.

LEFT: A sandhill crane makes its graceful entrance. The sounds of these birds returning to Montana in the spring will always cause you to crank your head upward in amazement.

RIGHT: This golden eagle is dining on fresh ground squirrel. Unlike the bald eagle, which feeds primarily on fish, the golden eagle makes a meal of small mammals, game birds, and snakes—as well as carrion.

BELOW: Of the turkey subspecies found in the United States, the Merriam's is what you will find through most of Montana; however the eastern subspecies is quite common in the upper Flathead Valley.

FACING PAGE: This mule deer buck is bounding, or "stotting"—traveling with all four feet off the ground. This better enables them to escape predators by jumping over obstacles that pursuers may have to go around.

RIGHT: Mother northern flicker addresses a hungry batch of young on her return to the nest. Once the young can stick their heads out of the nest, it is usually only a matter of days before they fledge.

BELOW: The aspen bark is tasty, but a wary beaver eyes the camera while eating.

FACING PAGE: A bull elk reduces a spruce tree to a stick. Seldom have I watched a bull leave a rub before completely destroying the tree. At this time, he becomes vulnerable to having his cows stolen by another bull because of his preoccupation with rubbing the tree.

LEFT: "Heckle and Jeckle": two common ravens atop a spruce. One of Montana's hardiest birds, ravens withstand the harshest of conditions, and can make a meal out of just about anything.

BELOW: A baby mountain lion peers from a crack within its den located deep in the Rattlesnake Wilderness Area of western Montana.

FACING PAGE: This large male grizzly bear is on the move. It's late May and it's breeding season, so he has much territory to cover. I found this same fellow fifteen miles away a day and a half later.

It's not all head-butting! When one male gets too close to another's ewes, the chase is on. In this case, the pursuer pushed the pursued down an entire hundred-foot embankment.

RIGHT: Found at high elevations, this whitetailed ptarmigan with its snow-white winter plumage will bury itself in the snow at night to avoid detection.

BELOW: A small meal but still a meal: with temperatures well below zero, this otter makes good use of a hole in the ice to retrieve small cutthroat trout.

It's mid-August, and this whitetailed buck is sporting his new set of antlers.
Within the next two weeks, the velvet will give way to hard-surfaced antler.

Red fox kits ham it up for the camera. I had to hide in a blind to film them, but sometimes the whirl from the motor-drive would cause everyone to disappear into the nearest hole, while other times I was rewarded with a peculiar stare.

ABOVE: Baby porcupines are one of the largest newborn mammals in comparison to their mothers.

RIGHT: It's mid-October and, with the rut over, this bull elk will spend a good deal of his time eating, replenishing his fat reserves before the onset of winter.

Autumn is turning to winter, so this cow moose probably will join a small group
of her peers until spring, when they again separate to calve.

Curiosity has slowed down these newborn bighorn lambs long enough for me to get a few portraits. For most baby animals, playing is top priority— and for these lambs, there's no exception.

ABOVE: An April snow squall along the Rocky Mountain Front catches a short-eared owl in a cattail marsh.

FACING PAGE: A whitetailed deer buck stands chest deep in the snow during the winter of 1996-1997, when snow came early and stayed long. Chances are this buck did not make it, since it was photographed in an area where up to 70% of the deer were lost to winter-kill.

LEFT: A small bull elk follows his harem across a river. It's the first week of October and most of these cows have probably already been bred, which is why this bull is not taunted by a larger bull.

BELOW: Mountain bluebirds like this male are a common sight along rural roads throughout Montana.

I was photographing some bighorn sheep when this black wolf came along and caught their attention. The sheep stood and stared with great intensity until the wolf disappeared into the forest, at which time they all turned their backs and bedded down.

RIGHT: Hovering against a winter sky, a northern shrike searches a hedgerow for any movement. This shrike is only a winter visitor in Montana, while its cousin, the loggerhead shrike, is a summer resident.

BELOW: A small group of cow and calf elk pause on the open prairie during a cold November morning.

RIGHT: Yellow iris and cattails attract a male yellow-headed blackbird.

BELOW: A common loon, with chicks that once had been small enough for both to be carried on her back, waits patiently as her mate searches below for a meal of baby northern pike.

FACING PAGE: A cinnamon black bear perched precariously in a cottonwood tree peers down on me as if to ask for help. Actually, this bear is feeding on the cottonwood fruit, called catkins. This may result in a lot of damage to the tree, but generally not enough to kill the tree.

First-year whitetailed deer fawn and a mature buck that is four to six years old.

ABOVE: Backlit snow geese lift off against the morning sky at Freezeout Lake along the Rocky Mountain Front.

RIGHT A badger emerges from his new excavation a bit dusty. The low-rider of the weasel family, badgers can be found all around Montana.

89

ABOVE: A Canada goose baby awakes on its first-ever morning to find frost, but

ABOVE: Out of the river, a grizzly bear delivers a shower—much like a Labrador after a retrieve.

RIGHT: A sow and a 2½ year old cub sit survey the masses below on opening day of Going-to-the-Sun Road in Glacier National Park. What was yesterday a quiet path, is now a multicolored noisemaker.

Like kangaroos,
two cow elk box
it out along the
river's edge.

LEFT: A male lazuli bunting fills the air with song on a summer's morning. These birds are more often heard than seen so take binoculars in hand to seek them on the hillsides surrounding Missoula and in the Flathead Valley.

BELOW: Like the wolf, the lynx is a subject rarely filmed in the wild. As luck would have it, my turn came when this one crossed the road. When filming only wild subjects, one must take what is given.

ABOVE: Just as house cats do, wild felines entertain a certain amount of curiosity. This sub-adult mountain lion paralleled the photographer along a hiking trail in Glacier National Park for a short time, then left. Stories like this are common, but in most cases a camera wasn't available.

FACING PAGE: Namesake residents of the National Bison Range, with the Mission Mountains beyond.

ABOVE: Locked on its prey, an osprey hovers. Twenty-five years ago, sighting these birds along Montana rivers was less common than it is today, but after DDT was outlawed, these birds have more than rebounded. Now, finding ospreys' notable nests is commonplace.

LEFT: A mountain goat surveys his domain from the slopes overlooking Hidden Lake in Glacier National Park.

LEFT: A swift fox dad shelters his young pup. Once eliminated along Montana's Front Range prairie, this small fox has been successfully reintroduced, especially on the Blackfeet Indian Reservation east of Glacier National Park.

BELOW: At first light, this whitetailed deer feeds along a small opening in the forest.

With the rut winding down, bull elk energy reserves dwindle and activity levels fall, so this individual bugles from his bed in evening's last light. I've referred to viewing the end of the rut as like stepping into a boxing arena to watch the last couple of rounds, when injuries are evident and heads are often hanging low.

A great blue heron in full breeding plumage gives the appearance of a
Concorde jet. The great blue can be found throughout Montana wherever
there is water. Although fish is its primary food, it will eat crawfish, snakes, or
even mice, stabbed with its long bill.

Being very social and expressive animals, coyotes use their howls—as on this cold February morning—

LEFT: Kinnikinnick supplies a bed for this red squirrel as it works on a pine nut.

BELOW: Reclusive in the first two months of its life, a whitetailed deer fawn peers carefully around. At this age, it will associate only with its siblings and mother.

FACING PAGE: Seeming to mimick a monkey, this brown-phase black bear cub holds onto a small lodgepole tree. Though called "black," the species appears in a variety of color phases—including brown, cinnamon, blond and even blue (in the glacier bears of southeast Alaska).

ABOVE: Occasionally common redpolls, like this one sitting out a snowstorm on a sunflower stalk, invade Montana during winter. During the winter of 2001-2002, flocks numbering in the hundreds could be found visiting bird feeders.

FACING PAGE: With snow on the ground, this bull moose's time for wearing antlers is short. By mid-December he probably will have shed the antlers, leaving bare pedestals where they had been attached.

A bull elk may weighs 800 pounds or more, with the largest topping the scales above 1,000 pounds. It generally takes an elk at least four years to develop a rack such as this one.

RIGHT: Hustling back to its burrow with a mouth full of grass, this black-tailed prairie dog represents a community that offers sanctuary to many animals, such as rattlesnakes, burrowing owls, and Montana's rarest mammal, the black-footed ferret.

BELOW: A very large male grizzly races down an embankment in hot pursuit of a female. But she was so intent on getting away from this particular fellow that she ran down the road where I was parked, and sat in front of my vehicle until the male disappeared. She then fed on the road's shoulder near the vehicle, all the while keeping her eyes on the area where the male had disappeared.

What a fine back-scratcher! Early each year, black bears like this one leave their dens with long winter coats. But, as spring progresses, those thick coats become increasingly uncomfortable and itchy, making trees like this spruce into convenient tools of choice for the bears' comfort.

A young mule deer buck drinks from a temporary pond of snowmelt
high upon Logan Pass in Glacier National Park.

RIGHT: Its black mask appropriately gives this raccoon the appearance of a bandit, which the creature sometimes is. The species has done quite well in Montana, ever expanding its range, often in urban areas where it very easily adapts.

BELOW: A desert cottontail sprints across the open prairie of the Charles M. Russell National Wildlife Refuge.

Coyotes hunt alone, in pairs, or in small groups, but this single guy on the hunt pauses to peer into the camera through a tangle of willows.

Always on guard, a bull elk cranks his head upward. He is a royal bull, hunters' term for an elk with six points on each antler, so he carries, on average, more than twenty pounds of headgear.

RIGHT: The western meadowlark, Montana's state bird, gives out the first sounds of spring on the prairie.

BELOW: A baby, or "kid," mountain goat stretches its tiny frame after arising from a nap. It will follow mother over some of Montana's steepest ground. During those travels, the mother will always keep it on her uphill side for safety.

Two bull elk position for a duel, which consisted mostly of posturing and stare-downs, with only an occasional clash of their antlers.

ABOVE: Merry Christmoose!

RIGHT: A grizzly bear makes his way along the edge of a lodgepole forest. There are very few areas left in the Lower 48 States that support grizzly bear populations, Montana being one of the strongest. The grizzly is the king of beasts here, and a great symbol of wilderness, which would seem much emptier without him.

A bull elk crests a ridge through a full moon. To me, this picture conveys
several things: space, beauty, wildness, and peace. I find all these things
to be symbolic of where I live: Montana.